Fitsum Melaku

Domestic Solid Waste Management

Assessment of Management Options for Domestic Solid Waste in Addis Ababa Case Study of French Legasion Area

VDM Verlag Dr. Müller

Impressum/Imprint (nur für Deutschland/ only for Germany)
Bibliografische Information der Deutschen Nationalbibliothek: Die Deutsche Nationalbibliothek verzeichnet diese Publikation in der Deutschen Nationalbibliografie; detaillierte bibliografische Daten sind im Internet über http://dnb.d-nb.de abrufbar.
Alle in diesem Buch genannten Marken und Produktnamen unterliegen warenzeichen-, marken- oder patentrechtlichem Schutz bzw. sind Warenzeichen oder eingetragene Warenzeichen der jeweiligen Inhaber. Die Wiedergabe von Marken, Produktnamen, Gebrauchsnamen, Handelsnamen, Warenbezeichnungen u.s.w. in diesem Werk berechtigt auch ohne besondere Kennzeichnung nicht zu der Annahme, dass solche Namen im Sinne der Warenzeichen- und Markenschutzgesetzgebung als frei zu betrachten wären und daher von jedermann benutzt werden dürften.

Coverbild: www.ingimage.com

Verlag: VDM Verlag Dr. Müller Aktiengesellschaft & Co. KG
Dudweiler Landstr. 99, 66123 Saarbrücken, Deutschland
Telefon +49 681 9100-698, Telefax +49 681 9100-988
Email: info@vdm-verlag.de

Herstellung in Deutschland:
Schaltungsdienst Lange o.H.G., Berlin
Books on Demand GmbH, Norderstedt
Reha GmbH, Saarbrücken
Amazon Distribution GmbH, Leipzig
ISBN: 978-3-639-25749-6

Imprint (only for USA, GB)
Bibliographic information published by the Deutsche Nationalbibliothek: The Deutsche Nationalbibliothek lists this publication in the Deutsche Nationalbibliografie; detailed bibliographic data are available in the Internet at http://dnb.d-nb.de.
Any brand names and product names mentioned in this book are subject to trademark, brand or patent protection and are trademarks or registered trademarks of their respective holders. The use of brand names, product names, common names, trade names, product descriptions etc. even without a particular marking in this works is in no way to be construed to mean that such names may be regarded as unrestricted in respect of trademark and brand protection legislation and could thus be used by anyone.

Cover image: www.ingimage.com

Publisher: VDM Verlag Dr. Müller Aktiengesellschaft & Co. KG
Dudweiler Landstr. 99, 66123 Saarbrücken, Germany
Phone +49 681 9100-698, Fax +49 681 9100-988
Email: info@vdm-publishing.com

Printed in the U.S.A.
Printed in the U.K. by (see last page)
ISBN: 978-3-639-25749-6

TABLE OF CONTENTS

i

LIST OF TABLES

LIST OF PLATES

ACRONYMS

AAU	Addis Ababa University
AASBPDA	Addis Ababa Sanitation, Beautification and Park Development Agency
AAWSSA	Addis Ababa Water Supply and Sewerage Authority
EHD	Environmental Health Department
FEPA	Federal Environmental protection Authority
IMWM	Integrated Municipal Waste Management
ILRI	International Livestock Research Center
ISWM	Integrated Solid Waste management
MSE	Micro/small Scale Enterprises
MSW	Municipal Solid Waste
NUPI	National Urban Planning Institute
SW	Solid Waste
SWM	Solid Waste Management
UNEP	United Nation Environmental Program
UNHCS	United Nations Center for Human Settlements
VCM	Volatile Combustible Matter

ABSTRACT

In any solid waste management study information on solid waste characteristics and rate of generation is essential. One of the main constraints for proper planning and design of solid waste management in Addis Ababa is the absence of reliable and up to date study in this area. Since the lion share of the municipal solid waste (76%) is contributed from residential houses giving solution to this part will contribute much for the over all management of the municipality solid waste.

The assessment of management options for domestic solid waste was done in an area called French Legasion situated in Yeka sub-city, one of the 10 sub-cities of Addis Ababa city government, with the objective of assessing the characteristics of solid waste generated at the household level and its level of impact on the environment as well as identifying sustainable management options accounting for economical, environmental feasibility and social attitudes.

To get reliable data 120 households were randomly selected from the study area. A structured questionnaire was used to collect household level data on the socio-economic and daily traits. The solid waste of each household was collected, sorted and weighted for 11 consecutive days for compositional and generation rate analysis. The proximate and ultimate analysis was also undertaken in the Addis Ababa chemical engineering laboratory and in the analytical laboratory of ILRI. Statistical software called Statistical Package for Social Science (SPSS 10) was used during the analysis for both the structured questionnaire and the collected solid waste data.

According to the data analysis the composition of the waste was found high in food waste (55.35%) followed by ash and dirt (22.29%); the average generation rate was determined to be 196 gm/cap/day with the average density of 311 kg/m^3. The correlation analysis clearly showed that the generation rate positively correlated to the socio-economic background of the households and negatively correlated to the educational background.

The survey analysis also showed that illegal solid waste disposal in open spaces is common practice in the area and some of the main reasons raised were the inappropriate placement and insufficient number of transfer stations and communal containers.

The result of the proximate analysis showed high moisture content and this directly related to the trend of high consumption of fresh vegetables. And this ultimately lowers the calorific value of the material (6001.27KJ/Kg). Incineration is not the best available option in the management of domestic solid waste due to the high capital cost, skilled manpower requirement and environmental burden.

The compositional analysis as well as the result of proximate and ultimate analysis clearly showed that the materials generated from residential houses are ideal for the preparation of compost, which ultimately diverts organics from entering the MSW stream. Since organics are the largest components of residential solid waste, the greatest reduction in waste collection and disposal can be achieved by diverting this component of the waste stream.

In order to attain a sustainable management of solid waste much effort has to be done on educating of peoples towards the management options; and much efforts has to be forwarded for individuals and enterprises engaged on the recovery of waste.

I. INTRODUCTION

1.1 BACKGROUND

Addis Ababa the capital city of Ethiopia is the largest as well as the dominant political, economical, cultural and historical city of the country. The city generates large volume of waste from households, public institutions and industries. Among all waste, solid waste is a major visible source of environmental pollution in the city, which in turn, is the cause of pollution of air, water and soil. As a result, it is seriously affecting human health, the quality of life and natural resources. Thus the solid waste littering the city is becoming one of the major areas of apprehension.

With the current growth rate of urban population in Ethiopia, it is estimated that the population of most urban areas is doubling every 15-25 years. As solid waste generation increases with economic development and population growth, the amount of solid waste generated in the cities and towns will be expected to double in an even shorter time span. Hence the city's solid waste management cost will increase accordingly (Yami, 1999).

Although solid waste is the serious problem of Addis Ababa city there is limited research made on solid waste management. According to the Addis Ababa Sanitation, Beautification and Park Development Agency (AASBPDA) sectoral plan document the collection and disposal of solid waste is stated as a persistent problem of the city. In the document, the current solid waste generation / day in the city is estimated to be $2,237m^3$ out of which about 65% is being collected and disposed in controlled disposal sites. The remaining large proportion of the solid waste is littering open spaces, ditches and rivers near the source of waste generation (AASBPDA, 2004).

In most western cities the cost of municipal solid waste management is between 20 -50 % of the total municipal expenditure (AACC, 2002). Taking in to account the budget allocated for the Addis Ababa city solid waste management is only 1 % of the total city government budget [AASBPDA, 2004], this makes the estimated 65% collection efficiency of the city questionable.

1

According to the AASBPDA report on current status of solid waste management the available studies show low waste production rate estimated to be 0.15 to 0.252 kg/capital/day. The density is usually high, as it varies from 205 - 370 kg/m^3 and average being 333 kg/m^3, the measurements were done some ten years ago and it cannot show the present picture and requires current study (AASBPDA, 2004).

Studies showed about 76% of the solid waste of the city is generated from households, 6% from street sweeping, 9% from commercial facilities, 5% from industries, 3% from hotels, and 1% from hospital (AASBPDA, 2004). This can clearly indicates that households take the major share of solid waste generated in the city. In this study we focused on the major solid waste source: domestic solid waste.

In any solid waste management study information on solid waste characteristics and rate of generation is essential. The characteristics dictate the type of management technology to be used while the rate of generation is used for sizing and economic analysis.

There are limited researches made on the socio-economic condition of the population of Addis Ababa. The 1980 Addis Ababa household social survey stated that low income (57%), middle income (35%) and high income (8%) of the population earns less than 100,100-400, and more than 400 Birr per month respectively (AACG, 1980). The 1993 Addis Ababa wastewater plan study claimed that the low income, middle income and high income as 91.5%, 5.6% and 2.9% respectively as a function of housing density (AAWSSA, 1993). In 1997 the Region 14 administration claims that the low income (60%), middle income (25%) and high income (15%) groups earns less than 300, 300-600 and more than 600 Birr per month respectively (AACG, 1997).

For this study the 1997 Region 14 administration socio-economic studies on income classification was adapted But since the purchasing power of money depreciates with time this study incorporates the inflation rate and brings the net present value of the money to the study year. The results obtained outlined in the following **Table 1.1**.

2

Table 1.1 Income classifications

Income category	Income per adult per month for the year 1997	Deflated income per adult per month
Low income	< 300	< 337
Middle income	300 - 600	337 – 673.8
High income	>600	> 673.8

1US Dollar = 9.22 Ethiopian Birr

In this study domestic solid waste generation rate and composition analysis were conducted and then the potential management options were critically assessed taking into consideration their environmental, economical and social feasibility to the area under consideration.

The study area is selected for the following reasons:

- It is an appropriate place to determine the domestic solid waste generation rate and composition since more than 90% of the area is under residential houses.
- The economic background of the population around the area is relatively low such that the proposed solid waste management options may create job opportunities.
- Compared to other areas of Addis Ababa there is relatively sufficient free space for the implementation of the proposed solid waste management options.
- Due to the up and down topographic nature and poor settlement pattern some areas are inaccessible for trucks in addition to the insufficient service of the municipality: the problem of solid waste is more significant.

Organizations working on the solid waste management will benefit a lot from this thesis. Some of these are:

✓ Addis Ababa sanitation, Beautification and Park development Agency.

✓ NGOs participating in the environmental protection.

✓ Industries and Private investors will get some insight on the extent of the economic potential of domestic solid waste management option.

✓ The Results will also be used for planning further research in solid waste management.

1.2 OBJECTIVE

1.2.1 GENERAL OBJECTIVE

- To assess the characteristics of solid waste generated at the household level and its level of impact on the environment.

- To identify sustainable management options accounting for economic, environmental feasibility and societal attitude.

1.2.2 SPECIFIC OBJECTIVE

- Estimating the amount of solid waste generated per capital per day for the selected area.

- Characterizing solid waste generated at the household level.

- Assessing the impact of household solid waste on the environment.

- Identifying sustainable management options.

- Assessing environmental and economic feasibility and societal attitude of the proposed management options.

II. LITRATURE REVIEW

2.1 General condition of solid waste

Solid waste is material, which is not in liquid form, and has no value to the persons who is responsible for it (Chris Zurbrugg, 2003), from the days of primitive society, humans and animals have used the resources of the earth to support life and to dispose of waste .In early times, the disposal of human and other waste did not pose a significant problem, for the population was small and the amount of land available for the assimilation of waste was large. Problems with the disposal of waste can be traced from the time when humans first began to congregate in tribes, villages, and communities and the accumulation of waste become a consequence of life (Techobanaglous et al., 1993).

The booming growth of cities of the developing world has outpaced the financial and manpower resources of municipalities to deal with provision and management of services, of which solid waste is the major one. Lack of these services greatly affects the urban poor, women and children who are vulnerable to health hazards (Yami Birke, 1999). Improper management of solid waste has direct adverse effects on health. Uncontrolled fermentation of garbage creates a food source and habitat for bacterial growth. In the same environment, insects, rodents, and some birds' species proliferate and act as passive vectors in the transmission of some infectious diseases (Gerald, 1997). About 22 human diseases are related to improper solid waste management (World Bank, 1999).

To a large extent ecological phenomena such as water and air pollution have also been attributed to improper management of solid waste (Techobanaglous et al., 1993).

2.2 Source, Composition and Generation of solid waste

Knowledge of the source and types of solid waste, along with data on the composition and rate of generation, is basic to the design and operation of the functional elements associated with the management of solid waste (Techobanaglous et al., 1993).

6

According to Techobanaglous et al., (1993) the sources of solid waste categorized into:

- Residential
- Commercial
- Institutional
- Construction and demolition
- Municipal services
- Treatment plant sites
- Industrial and agricultural

The municipal solid waste (MSW) is normally assumed to include all community waste with the exception of industrial process waste and agricultural waste.

According to the Addis Ababa sanitation, beautification and park development agency (AASBPDA), 2004 the source of solid waste in Addis is assumed 76% from household, 18% from institution and 6% from street sweeping.

Composition is the term used to describe the individual components that make up a solid waste stream and their relative distribution, usually based on percent by weight (Gerald, 1997). Information on the composition of solid waste is important in evaluating equipment needs, systems, and management programs and plans (Gerald, 1997).

The quality of solid waste in developing countries is quite poor in comparison with that found in industrialized countries. This is an important factor to bear in mind when treatment and recycling programs are being considered as economic options (Jorge, 2003).

Based on the study carried out by Louise Burger (1995) the percentage composition by weight for combustible materials (leaves, grass, etc) is estimated to be about 22%, for non-combustible (stone, etc) 3%, for fines greater than 10mm size (food waste, straws, etc) 34%, for fines (ashes) less than 10mm size 28% and for recyclable materials (paper, wood, metals, plastics, rubber, etc) 13%.

Literatures reviewed reveals that waste generation rates are affected by degree of industrialization, climate and socio-economic development. As economic prosperity increases, the amount of solid waste produced consists mostly of luxury waste such as paper, cardboard, plastic and heavier organic materials. Some literatures show that in cities of developing countries waste densities and moisture contents are much higher (Fitsum Haile, 2003). Knowledge of the amount of solid waste generation is necessary to design management strategies to effectively handle those wastes (Techobanaglous et al., 1993).

Table 2.1 Waste generation rate

Solid waste	Countries		
	Low income	Middle income	Industrialized
Production Kg/cap/day	0.3 - 0.6	0.5 - 1.0	0.7 - 2.2
Ton/cap/year	0.2	0.3	0.6

Source: Urban Management Program – UNEP. Regional Office for Latin America and the Caribbean. Solid Waste/ Private Sector and Sanitary Landfills. Urban Management Series. Vol 13.(undated)

Regarding the solid waste generation rate of Addis Ababa there is no up-to-date and reliable data. There are few studies undertaken since 1982 by different bodies. Nur consult carried out the first main study in 1982. The result suggested that per capital generation of solid waste was 0.15kg/day/person with a 1% annual growth rate and approximate waste density of 370 kg/m^3 (Zerayakob, 2002). Louise Burger international in 1986 carried out the second study and the result was a rate of 0.20kg/day/person. Louise Burger international also carried out the third available study in 1994 and 1995. According to the 1994 study, the average per capital generation of solid waste was 0.22kg/day/person and the density about 336 kg / m^3. The 1994 and 1995 study shows the generation dependency on income levels. Based on the data collected in 1994, generation/capital/day was 0.35kg, 0.28kg, and 0.17kg, for high, medium, low-income groups respectively. The data of 1995 collected shows 0.47kg, 0.236kg, and 0.26kg for high, medium and low respectively (Zerayakob, 2002).

The last available data studied at sub-city level by Yitayal (2005) in his quantity and compositional analysis he randomly selected 197 households from the Arada sub-city and come up with a result of 116.9, 153 and 162.5 gm/cap/day for low income, middle income, and high income groups with average density of 159.8 Kg/m³.

Compared with the former studies, which were made on sampling at the disposal containers, the later methodology seemed reliable since it took into account the household size and income status of the households and collected sample at the sources. However the figures of the later study shows much deviation from the former studies and other similar studies made in developing countries. Thus, it seems there are other factors, which need to be looked into.

2.3 Methods of Solid waste characterization

Four methods for estimating waste quantities and composition can be identified: direct sampling (also referred to as waste stream analysis and waste audits), material flow, surveying waste generators, and literature sources (Gerald, 1997).

Direct sampling: Direct sampling involves sampling, sorting, and weighing materials from the waste stream of a specific generator. This method has been used to estimate the composition of municipal waste streams. Representative sampling methods must be employed to achieve accurate results. When using the direct sampling method, the following questions must be addressed: How will representative samples of waste be obtained; and how many samples should be selected to achieve the desired level of accuracy in the results? The responses to these questions will influence the cost of conducting the study as well as the usefulness of the data.

Waste stream analysis: Waste stream analysis is another term used for characterizing the waste stream of a specific operation for a designated time period. Waste stream analysis is defined as a method for collecting, sorting, and measuring the amount and type of waste generated by an operation. Results of a waste stream analysis provide data about the amount and type of waste/residues in the waste stream. Data should be collected for a minimum of one week; the length of time depends on how the data are to be used and the accuracy required. The results are averaged to estimate the amount of waste that the facility generates for a period of time.

Waste audit: The basic objectives of a waste audit are similar to a waste stream analysis. A waste audit involves a more detailed assessment of waste. The waste audit assesses not only the output (waste), but also the input, such as food products, packaging materials, office supplies, mail, or any process that results in materials that must be discarded. The detailed and complicated analysis of material flow through an institution will enable the facility to find the amount purchased, used, recycled, and disposed of for different materials. A waste audit can involve all materials or focus on a specific material, such as cardboard or office paper that is generated by a facility or department.

Material flow: The material flow method applies the concept of conservation of mass to track quantities of materials as they move through a defined system or region. The material flow methodology in this instance is based on the production weight data for materials and products. Generation data are the result of making specific adjustments for imports, exports, and diversions to the production data by each material and product category. The method also considers the useful life of products. One of the problems with the material flow approach is that it is difficult to quantify product residues, such as food left in the container and detergent remaining in the package.

Surveying waste: Surveying industrial generators, such as food processors can provide useful data in quantifying waste generation. More accurate data can be obtained if the waste/residues are measured at the disposal site.

Literature sources: Data on waste/residues quantities and composition are available from a variety of sources including public agency documents, engineering reports, trade publications, and professional journals. These data may be helpful in assisting managers in identifying the type of residues/waste generated by a specific industry or activity. However, caution should be exercised when operational decisions are made based on data from the secondary sources. . Waste characterization and generation rate studies are recommended for operational uses rather than relying on published data since each study site is unique.

2.4 Evolution of solid waste management

Waste management in general and solid waste management in particular evolved from a practice which assumes the environment to have an infinite waste assimilative capacity to that which recognizes the limits of the environment. **Table 2.2** represents the chronology of dominant waste management practices in different decades.

Table 2.2 Chronology of waste management practices in different decades

Decade	Waste management practices
1960	Dilution is the solution for pollution: all the generated waste is dumped into the environment without any treatment.
1970	Treatment: waste is treated to the required level before discharging/dumping it into the environment. But the residual of the treatment poses another problem.
1980	Recycling: some part of the waste is used as resource as such the amount waste to be disposed decreases.
1990-2000+	Source reduction and Green design: these strategies emphasize waste prevention and minimization.

Source: Helge B. (1997)

2.5 Components of solid waste management

Solid Waste Management (SWM) broadly, refers to the material flow stream of waste from generation to ultimate disposal and comprises storage, collection, transportation/ transfer, processing (reuse/recycling/ composting), and disposal.

2.5.1 Storage

The size of premises, nature (type) and generation rate of solid waste determines the type of storage to be used. Storage facilities must be animal and insect proof washable and robust enough to meet the exigencies in normal use. There is a limit to the duration that solid waste can be stored at source (in the premises) based on the type and source of solid waste.

Solid waste should be collected and disposed of from temporary stores to final disposal site before breeding various disease-carrying vectors. Uncovered containers of waste are exposed to

11

human and animal scavengers that litter waste around and create community health problems.

2.5.2 Collection and transportation /transfer

Collection refers to the art of removing accumulated waste, be it containerized or not, from generating sources. Collection may occur at a centralized location where generators deliver their solid waste or by going from individual generator to another, which increases the expense of collection. Transfer or transportation, as the name indicates, refers to the transportation or/and haul of solid waste from a central point to one or more distant final management facility (Gerald, 1997).

The level of service for waste collection varies markedly in industrialized and developing countries. In the former, services have expanded to the extent that over 90 per cent of the population (and 100 per cent of the urban population) has access to waste collection.

One of the most costly urban services in developing countries is solid waste management absorbing 20-40 percent of municipal revenue (Peter, 1996). Large portion of this cost is spent on collection of the solid waste generated. In spite of this, up to 30-60% of waste generated is not collected and less than 50 percent of the population is served (World Bank, 2006). Yet, it should be noted that municipal services in developing countries are handicapped by limited finances and an ever-increasing demand on urban services. The failure to provide adequate collection services poses a serious threat to human health. Solid waste collection and transportation system in developed countries constituted about 70-80 percent. In the USA, collection costs in 1986 were $10.4billion dollars out of $13.8billion spent on municipal waste costs (Murray, 2007). Minimizing collection and transportation cost as well as maintaining adequate service provision is one component of an efficient solid waste management (Mir Anjum Altaf, 1996).

According to the AASBPDA (2003) there are three modes of collection systems; communal container collection using lift and compacting trucks, institutional collection using lift trucks and door to door collection using compactor and side loader truck. Container collection covers 67% and door-to-door 33% of the total collection service given. There are about 150 micro and small enterprises, (MSE) which are engaged in pre-collection of solid waste.

12

In most part of the city, the primary solid waste collectors and households themselves deliver their waste to containers provided by municipality on the roadside and vacant plots. There are 500 larger 8m^3, which are metallic, open, and transported by lift trucks/volvo26, Nissan10/ in designated areas. There are also 476 smaller 1.1m^3 wheeled waste storage containers with lids where the waste is emptied by 9 Renault compactors. The 1.1-m^3 containers stand by the side of the roadway.

Households provided with house-to-house collection service empty their waste receptacles to 16 Nissan side loader truck and 10 Hino compactors. The agency claims to collect once or twice a week but as seen from observations the collection hours are not regular and the time interval between collection cycle seems longer.

Table 2.3 Collection and Transporting Equipment and Vehicles

No	Sub city	Population	Number of Containers		Number of Vehicles					Total	Iveco Route pecker truck	Grand Total
			8m^3	1.1m^3	Volvo /container lifter truck/ 8m^3	Nissan /container lifter truck/ 8m^3	Nissan side loader /10m^3/	Renault compactor /1.1m^3 container lifter/	Hino compactor /Kuka rotating truck/			
1.	Arada	315355	79	96	3	1	2	1	1	9	use with Gulele	
2.	Addis Ketema	332564	51	13	4	1	2	1	1	8	use with Kolfe	
3.	Lideta	307324	82	63	2	1	1	1	1	6	use with chrekers	
4.	Cherkos	318078	92	120	3	1	2	1	1	8		
5.	Yeka	308338	30	26	3	1	2	1	1	7	use with Bole	
6.	Bole	285198	38	65	3	1	1	1	1	8		
7.	Akaki Kaliti	180498	25	--	2	1	1	--	1	7	use with Nifas Silk	
8.	Nifas Silk	356498	44	49	3	1	1	1	1	5		
9.	Kolfe Keraniyo	271112	33	15	2	1	1	1	1	7		
10.	Gulele	330611	26	29	3	1	2	1	1	7		
	Total	3035138	500	476	28	10	15	9	10	72	5	77

Source: AASPDA, 2004

The municipal solid waste management is mainly financed from the general revenue of the city government. According to the agency the budget allocated to solid waste management accounts for only 1% of the total city government budget.

Table 2.4 Municipal budgets for AASPDA from year 1995-2003

Year	1995	1996	1997	1998	1999	2000	2001	2002	2003	
									Treasury	Treasury + Donor
Million Birr for MSWM	4.1	4.9	N/A	N/A	N/A	8	10	9	6.5 (0.6%)	55 (3.6%)

Source: AASPDA, 2004

Table 2.5 Budget allocated for Yeka sub-city and to SWM sector

Budget Year	Total budget of the Yeka sub-city (Birr)	Budget allocated for SWM sector (Birr)	Cost of collection (Birr)
2004/05	45,251,400	1,317,200	260,100
2005/06	67,086,700	1,928,300	440,500
2006/07	68,695,400	1,885,000	444,000

As seen from **Table 2.5** for the last three consecutive years, the sub-city solid waste management sector took 2.91%, 2.87% and 2.74% of the total sub-city budget respectively. Taking in to account the 20-50 % of the total municipal budget allocated to solid waste management in other countries it can show us that the attention given to solid waste management is still minimal to provide reliable and sufficient service. From the solid waste management sector budget the cost of waste collection absorbs 19.74%, 22.84%and 23.55% of this share.

2.5.3 Reuse/ recycling and composting

Recycling/ reuse is the removal or diversion of material from solid waste discarded as useless and the use of material for the same purpose as was originally designed for, for other use in its

14

original form, or processing (treatment and reconstruction) of material to produce secondary row material for other products (Gerald, 1997).

In the industrialized countries recycling activities are widely practiced and are on the increase, primarily due to the political pressure of public opposition to disposal sites, and the economic pressure of the high cost of waste disposal attributable to land shortage, increasing costs of sanitary landfills, the unwillingness on the public's part to have landfills located in "their backyards", and stringent regulatory standards of waste disposal. In developing countries, on the other hand, which are still grappling with the basic task of collecting garbage, recycling of waste is carried out as a means of income generation.

Composting is the biological decomposition and stabilization of organic waste. It can be beneficial when applied on land. Composting operations of solid waste include preparing refuse and degrading organic matter by aerobic microorganisms. Refuse is presorted, to remove materials that might have salvage value or cannot be composted, and is ground up to improve the efficiency of the decomposition process. The refuse is placed in long piles on the ground or deposited in mechanical systems, where it is degraded biologically to humus with a total nitrogen, phosphorus, and potassium content of 1 to 3 percent, depending on the material being composted. After about three weeks, the product is ready for curing, blending with additives, bagging, and marketing.

In the Philippines a growing number of local governments are implementing ISWM, which includes waste reduction, recycling; composting and reuse estimates have shown that trades in waste materials has increased in volume by 39 %(World Bank, 2001).

Some of the key factors that affect the potential for resource recovery are the cost of the separated material, its purity, its quantity, and its location with regard to the intermediate and final processing facilities. The cost of storage and transport are the major factors that decide the economic potential for resource recovery (Chris Zurbrugg, 2003).

2.5.4 Thermal treatment-combustion/incineration

To reduce waste volume, local governments or private operators implement a controlled burning process called combustion or incineration. In addition to reducing volume, combustors, when properly equipped, can convert water into steam to fuel heating systems or generate electricity. Incineration facilities can also remove materials for recycling.

A variety of pollution control technologies significantly reduce the gases emitted into the air, including:

- Scrubbers—devices that use a liquid spray to neutralize acid gases
- Filters—remove tiny ash particles

Burning waste at extremely high temperatures also destroys chemical compounds and disease-causing bacteria. Regular testing ensures that residual ash is non-hazardous before being land filled (Chris Zurbrugg, 2003).

For centuries, burning has been a popular method of reducing the volume of solid waste. The burning of waste was rampant and uncontrolled. While uncontrolled burning of solid waste can be detrimental to health and the environment, confined and controlled burning, known as combustion, can not only decrease the volume of solid waste destined for landfills, but can also recover energy from the waste-burning process.

In many developing countries the domestic waste contain large amount of inert, such as sand, ash, dust and stones and high moisture level because of the high usage of fresh fruit and vegetables. These factors make the waste unsuitable for incineration (Chris Zurbrugg, 2003).

2.5.5 Solid waste disposal

Despite the effectiveness of source reduction, recycling, and combustion, there will always be waste that cannot be diverted from landfills. The safe and reliable long-term disposal of solid waste residue is an important component of integrated waste management. Solid waste residues are waste components that are not recycled, that remain after processing at a material recovery facility, or that remain after the recovery of conversion products and/or energy (Techobanaglous et al., 1993)

In many developed countries, burial in controlled landfills continues to be the most prevalent means of disposing of solid waste including hazardous waste. About 70% of the urban solid waste is disposed off in this way in the US and most European countries.

On the other hand most of the municipal solid waste (MSW) in developing countries is dumped on land in a more or less uncontrolled manner. These dumps make very uneconomical use of the available space, allow free access to waste pickers, animals and flies and often produce unpleasant and hazardous smoke from slow burning fries (Chris Zurbrugg, 2003)

Landfill gas is produced in landfill sites due to the anaerobic degradation of biodegradable organic waste. The gas produced is typically about 60% of methane and 40% CO_2. Landfill gas, with high content of methane, is potentially explosive and, as such, needs to be controlled. In some means of controlling (extracting) the gas is not used, the gas can migrate off site, causing problem to the surrounding environment (Gerald, 1997).

A survey made on 15 randomly selected large urban towns (Dessie, Bahir-Dar, Debrezaite, Gonder, Mekele, Nazareth) and medium urban areas (woldiya, Axum, Adigrat, Robe, Gimbi, Adwa, Arbaminch, Wolayita-Sodo, Debremarkos) shows that from the sample urban areas studied 13, that is 86.6% used open dump to dispose waste, while the rest used holes(Yami Birke,1999).

In Addis Ababa, more than 30% of the solid waste generated is dumped illegally to the environment, especially to the river, vacant lands, streets, drains, etc (AASBPDA, 2004).

Currently, there is one open dumpsite where a collected waste is disposed off. The site is getting full, surrounded by residential housing areas and public institutions and there is no daily soil cover (AASBPDA, 2004). The site is becoming detrimental to the surrounding environment. The city government acknowledges the dangers to the environment and the public health derived from the uncontrolled waste dumping. However, still the uncontrolled waste disposal is the best that is possible. Financial and institutional constraints are one of the main reasons for inadequate disposal of waste.

Table 2.6 Evolutions of waste disposal methods

Method of disposal	Description
Roadside disposal	This is common in areas where there is no waste collection service. The MSW is usually disposed of by its generators anywhere along the public highway or in a public dump
Uncontrolled waste disposal in small local dumps	There is a primary collection service and incipient transport to a nearby site where waste are disposed of without any control
Uncontrolled municipal dumping	There is primary and secondary collection. MSW is transferred and disposed of without control in a site on the outskirts of the city.
Controlled landfill	There is primary and secondary collection. MSW is transferred and disposed of with moderate control in a disposal site designed for the purpose and located on the outskirts of the city. The waste is buried regularly.
Sanitary landfill	The sanitary landfill is designed, built, and run according to sanitary and environmental engineering criteria. The site meets legal requirements and applies an environmental monitoring program. Environmental impacts are minimal and the population is not against the project.

Source: Adapted from "A Framework for the Disposal of Municipal Solid Waste in Developing Countries by Andrew Cotton, Mansoor Ali and Ken Westlake. Loughborough: WEDC, 1998.

2.6 Properties of solid waste

2.6.1 Physical property

According to Techobanaglous et al., 1993 the important physical properties of MSW include density (sometimes referred to as specific weight), moisture content, particle size and size distribution.

Density

This is the weight per unit volume and is expressed as kg/m³. Density varies because of the large variety of waste constituents, the degree of compaction, the state of decomposition, and in landfills because of the amount of daily cover and the total depth of waste.

Density is important because it is needed to assess the total mass and volume of waste, which must be managed. Density varies not only because of the type of treatment it gets (collection and compaction) but also because of geographical location, season, and length of time in storage. Some typical density values are presented in **Table 2.7**.

According to Louise Burger international, 1995 study the density of the Addis Ababa solid waste is estimated to vary from 205-370 kg/m3 with the average being 333 kg/m³.

Moisture content

The most commonly used method of expressing moisture content is as a percentage of the wet weight of material. Moisture content is important in regards to density, compaction, the role moisture plays in decomposition processes, the flushing of inorganic components, and the use of MSW in incinerators. The wet weight moisture content can be determined using the following equation:

$$M = \left(\frac{w-d}{w}\right)100$$

Where M = moisture content (%)

w = initial weight of sample (kg)

d = weight of sample after drying at 105°C (kg)

Some typical moisture contents are shown in **Table 2.8**

Table 2.7 Typical solid waste densities of different countries

Country	Waste density (kg/m3)
Industrial countries	
UK	150
USA	100
Middle income countries	
Egypt	330
Nigeria	250
Singapore	175
Tunisia	175
Low income countries	
Bangladesh	600
India	400-570
Indonesia	400
Pakistan	500
Thailand	250
Tanzania	330

Source: extracted from Fitsum Haile (2003)

Table 2.8 Typical Moisture Contents of Waste

Type of Waste	Moisture Content Range (%)	Moisture Content Typical (%)
Food waste (mixed)	50 - 80	70
Paper	4 - 10	6
Plastics	1 - 4	2
Yard Waste	30 - 80	60
Glass	1 - 4	2

Source: Tchobanoglous et al., (1993)

Particle size distribution

Particle size distribution, like the percentage of combustibles, is relevant to incineration and biological transformation methods. Particle size is also relevant for recycling and reuse and for equipment sizing for further treatment (Gerald, 1997). Particle size influences the bulk density, internal friction and flow characteristics, and drag forces of the materials (Tchobanoglous et al., 1977). Most vital of all, a reduced particle size increase the biochemical reaction rate during aerobic composting process and the most desirable particle size for composting is less than 2 inch (50 mm), but larger particles can be composted (Tchobanoglous et al., 1993). Similarly, for energy recovery, reduction of particle size is very important. For instance, for pyrolysis process, municipal refuse can be shredded to less than ½ in. (12.7 mm) size (Panovi et al., 1975).

2.6.2 Chemical property

Knowledge of the chemical composition of waste is important to help evaluate alternative processing and recovery options. This is especially important where waste are burned for energy recovery, in which case the three most important properties are proximate analysis, elemental analysis, and energy content. Elemental analysis is also important in determining nutrient availability.

Proximity analysis

Proximate analysis includes four tests - loss of moisture when heated to 105°C for 1 hour; volatile combustible matter (loss on ignition); fixed carbon; and ash (weight of residue after combustion).

Elemental analysis

This is also known as ultimate analysis and involves the determination of carbon, hydrogen, oxygen, nitrogen, sulphur, and ash. The results of this analysis are used to characterize the composition of the organic matter in waste. This is important for C:N ratios for biological decomposition. Sulphur is also important since during combustion it may contribute to environmental problems.

Energy content

The energy content of the components of waste can be determined using a boiler system, laboratory bomb calorimeter, or by calculation using elemental composition (Techobanaglous et al., 1993).

2.6.3 Biological property

Biological properties of MSW are relevant because of the technology of aerobic/anaerobic digestion to transform waste into energy and beneficial end products. Biodegradation can be aerobic or anaerobic. Anaerobic composting is the biological decomposition of 'food waste' with end products of methane, CO_2 and others. Some organic MSW components are undesirable for biological conversion, which is plastic, rubber, leather, and wood. The relevant fraction for biological transformation include fats, oils, proteins, Lignin, Cellulose, Hemi cellulose, Lignocellulose and Water-soluble constituents (Gerald, 1997)

22

III. MATERIALS AND METHODS

3.1 Description of the study area

The study area, French Legasion area, is situated in the Yeka sub-city, which is one of the ten sub-cities of Addis Ababa. Addis Ababa has a total area of 540 sq. km of which 18 sq. km is rural and situated between 2000-3000 meters above sea level. The Yeka sub city covers a total area of 59.07 sq. km with a population of 397,236 and situated in the northeast of Addis Ababa .It shares common boundaries in the northwest with the Oromia regional government, in the west with Gulale and Arada sub-city and in the southwest with Bole sub-city.

Regarding the climatic condition of Addis Ababa the lowest and the highest annual average temperature is about 10^0c and 25^0 c. The annual rainfall is around 1200 mm; the maximum occurs from June to September while the minimum occurs between March and May.

The study area, French Legasion, is a rapidly growing mainly residential area situated in the North East of Addis Ababa. Its population is estimated to be around 66,703 according to the Yeka sub-city administration. The housing compositions of the study area are presented in the following **Table 3.1.**

Table 3.1 Housing composition of the study area

Kebele	Number of houses		
	Self Owned residential	Government owned residential	Commercial houses
01/02	2047	949	225
03/04	3104	1107	314

Source: The administrative office of the respective kebele, 2006

The population size of the study area with sex distribution listed in the following **Table 3.2**.

Table 3.2 Population distributions with sex

Kebele	Population size		
	Male	*Female*	*Total*
01/02	15838	17762	32838
03/04	16255	17610	33865
Total	32093	35372	66703

3.2 Methodology

The methodology followed for the preparation of this study involves review of related literatures; preparation of the interview survey; establishment and training of crews for the collection and measurement of sample household daily solid waste and waste classification by type; processing of the survey data; analysis of the data and evaluation of findings; assessment of sustainable management options.

The followings are the major methodological steps followed during the study.

3.2.1 Preparation of a structured questionnaire

A structured, yet simple, questionnaire was designed to collect household level data on the socio-economic and daily waste traits. In addition the questionnaire included a number of attitudinal questions aimed at examining household awareness and attitudes towards the problem of urban solid waste. All the households randomly selected for this study filled the questionnaire prior to collection of samples.

3.2.2 Sample size determination

To determine the number of households that are going to be analyzed to obtain a reasonable and reliable result a method that was designed based on central limit theorem was used (Weiss, 1989) with a 95% confidence interval and a 1% error as a desired reliability. For sample size determination it is possible to use standard deviation of similar study such that a standard deviation of 0.056, which was determined in the Arada sub-city (Yitayal, 2005) was used.

According to the central limit theorem the size of sample determined using the following equation.

$$n = \frac{z^2 \delta^2}{SE^2}$$

Where Z= value of Z that corresponds to 95% confidence
Interval and equal to 1.96.

δ = Standard deviation

SE = Standard error

$$\text{Then, } n = \frac{(1.96)^2 (0.056)^2}{(0.01)^2}$$

$$= \quad 120$$

3.2.3 Identification of study households

In order to get a representative and reliable result, the total numbers of sampling houses that are going to be analyzed are determined from each Kebele based on their total number of houses. In the study area there are five kebeles 22, 07, 03, 04, and 06 after 2003 reform these kebeles grouped in to two administrative kebeles. But since the housing number of each former kebele is still valid the study was carried out based on former kebeles with a total number of houses of 522, 2135, 2405, 1566, and 584 respectively. From each Kebele the number of residential housing units was identified with their numbers and sampling units, as a proportion of total housing units were determined. Then for each Kebele the sample houses are selected based on their housing number using a simple random number table. After the house numbers of the housing units of each kebele randomly selected, the housing units were identified using their housing number.

3.2.4 Data collection

One day before collecting the samples from each household the plastic bags of different colors distributed for every household with the instruction that the organic waste (moist) should be kept in the red plastic bags and the non-organic (dry) ones in the yellow bags. And the identification number was assigned to each household and corresponding level is given for each and every bag distributed for each household.

On the next day early in the morning the collection of samples began. For the quality of the data the first day waste collected from each household was discarded taking into account that these waste may not be generated on a daily basis. Right after the second day up to the eleventh day (10 day) sample was collected on a daily basis.

Regarding the sampling process the crew was responsible for collecting the waste from each household and bringing to the place prepared for this purpose. Then after the waste collection, sorting of waste into different components was made for each household .The sorted components then weighted and their volume was determined using different sized wood boxes with a known volume. Finally the size distribution of the waste was determined using a 50mm and 10mm sized mesh wires and then the weight and volume measurements done for both size ranges.

3.2.5 Proximate analysis

The proximate analysis includes the determination of moisture content, volatile combustible matter (VCM), ash content and fixed carbon. In order to reduce the magnitude of error arising from the moisture change and from decomposition the analysis of the sample was started with in two to three hours after collection.

Care was also always taken to make the samples well mixed for this purpose each waste component were randomly taken and then chopped to reduce the size and then the well mixed sample finally was taken for laboratory analysis.

Regarding the determination of moisture content a sample from each component of the waste was daily taken to the chemical engineering lab just after the collection and analyzed using oven dry set at 105^0c for one hour (Techobanaglous et al., 1993).

The VCM and ash content determination was made at the last date of the collection program .The VCM was determined using a digital oven dry set at 950^0c for six minutes using a closed crucible. The temperature reaches 950^0c gradually in order to avoid flaming and protect the crucible from strong drafts to avoid mechanical loss of the specimen. The ash content of each component of waste was also determined using the same sample and equipment that was kept at 750^0c for three hours by using open crucibles (Techobanaglous et al., 1993).

Subtracting moisture content, ash and volatile matter from the initial sample determined fixed carbon.

3.2.6 Elemental (ultimate) analysis

The elemental analysis of waste components was done at the analytical lab of the International Livestock Research Institute (ILRI). The analysis involves the determination of nitrogen and carbon.

The determination of total nitrogen was carried out using the standard kjeldhal method. Regarding the determination of carbon due to the unavailability of analytical equipment and appropriate skill on the part of the analyst in developing countries UNEP, recommends a 'stop gap' approach suitable for composting in SWM is an estimation based on a formula developed in 1950s (Adams et al., 1951). The assumption is that for most biological materials the carbon content is between 45 to 60 percent of the volatile solids fraction. Assuming 55 percent is carbon.

The formula is as follows

$$\% \, Carbon = \frac{100 - \% \, Ash}{1.8}$$

And this study took this formula for the determination of carbon.

3.2.7 Data analysis

For the analysis of the sampled solid waste and survey questionnaire the statistical package for social studies (SPSS.10) was used.

In the data analysis the composition of waste was analyzed and per capital generation rate was determined.

In the correlation analysis the generation rate was cross-examined with the socio economic background and educational status.

28

3.3 Materials and Instruments

The following materials and instruments were used during the study:

A. Hand protective plastic gloves;
   - o   To protect hand from direct contact with dirt.
B. Mouth & Nose Mask;
   - o   To protect from bad smells and inhalation of any fumes.
C. Wood boxes (125cm^3, 1000cm^3, 3375 cm^{3}and 8000 cm^3)
   - o   For volume measurement
D. Balance scale
   - o   For weight measurement
E. A 50 mm and 10 mm wire mesh sorting metallic table
   - o   For particle size determination
F. Plastic sheets
   - o   To ensure no loss of waste during sorting
G.   Different type and color plastic bags
   - o   For the collection of solid waste from each household
H. Trash bags
   - o   For handling the collection of plastic bags
I. Camera
   - o   For capturing pictures of the working process
J. Bomb calorimeter
   - o   For the determination of heating value
K. Digital and non-digital oven
   - o   For proximate analysis (including moisture content, VCM and ash content)
L. Open and closed crucibles
   - o   Used during the determination of ash and VCM respectively

IV. RESULTS AND DISCUSSION

4.1 Survey results

Both survey data and observation indicate that self-occupied dwelling are the dominant in the area and renting houses to individuals is a norm in the study area. **Table 4.1** shows that 51.1 % of the households surveyed owned their houses followed by 44 % rented from the kebele and the rest rented from individuals.

Table 4.1 Housing characteristics

	Frequency	Percent	
Owned	62	51.7	
Rented from kebele	44	36.7	
Rented from individual	14	11.7	
Total	120	100.0	

Household size ranges from 1- 8 with the average of 4. The large family size is because of the extended family members. The average number of years of education of the most educated member of the household varied between 3rd grade and college degree. Out of the 120 households surveyed 51.7 % of the most educated member of the household varied between eight and twelve grade.

Table 4.2 Educational background of the most educated member of the household

	Frequency	Percent	
1 - 8	18	15.0	
8 -12	62	51.7	
>12	40	33.4	

The average monthly income of households per adult as computed from the survey data was Birr 146.89 for low-income groups, Birr 376.87 for middle-income groups and Birr 817.89 for high-income groups. The average monthly income of households per adult for the entire sample was Birr 240.47. In the following **Table 4.3** the socio-economic status of the households surveyed is presented.

Table 4.3 Income of households

Income group	No. Of households	Percentage (%)
Low income	84	70.58
Middle income	28	23.52
High income	7	5.88

4.2 Solid waste management in the study area

Like other areas of Addis Ababa the problem of solid waste is getting a little attention, but due to the economic inability of the households in the area to afford collection services, inappropriate placement and insufficient number of communal containers together with the area's undulating nature, the problem of solid waste littering around rivers and roadside is still one of the major environmental problems of the area.

According to the survey data, 58.3 % of the households are getting solid waste collection services by the primary collection schemes established by the youth and the rest of the households disposed their waste either on a communal container and/or open spaces.

Table 4.4 Households receiving a collection service

	Number of Households	Percent	
Have Collection	70	58.3	
No Collection	42	35.0	

Don't know	6	5.0
Not specified	2	1.7
	120	100.0

All easily transportable valuable waste materials (glass, tins, scrap metals etc) are collected and sold, by private collectors who visit households regularly for this purpose (commonly known as *kurali*), for waste recycler (individuals or companies). Thus, the actual volume of waste to be disposed of outside the house is decreasing and this reflected in the compositional analysis. High residential densities result in the generation of considerable amount of total waste in most neighborhoods. The waste consists primarily of organic matter from the kitchen which necessitates frequent disposal because of spoilage but most of the households getting the collection service only once in a week and this revealed in the disposal of some of the waste in the nearby open spaces.

Table 4.5 Frequencies of collection services

		Frequency	Number of Households	Percent	
Valid		Three times a week	17	14.2	
		Twice a week	2	1.7	
		Once a week	45	37.5	
		Less frequently	1	0.8	
		Don't know	40	33.3	

	Total	105	87.5
Missing	System	15	12.5
Total		120	100.0

Transport of waste from households is a growing problem. The city municipality employs communal collection points. However, the primary collectors or the households are responsible for transporting the waste to the collection point and the municipal transports the waste from the point to the ultimate disposal location, which is the 'Repie' landfill site.

Individual households who receive the collection service do not have standardized containers used to store waste prior to pick up. The survey analysis, as well as observation, shows it is up to the individual residences to designate some sort of collection container. Frequently, these are plastic bags or trash bags; however, the majority of the households simply place the trash bags full of waste on their gate to await collection.

Table 4.6 Containers for storing solid waste

	Item	Frequency	Percent
Valid	Metal or plastic container	21	17.5
	Basket or carton container	82	68.3
	No container	13	10.8
	Don't know	3	2.5
	Total	119	99.2
Missing	System	1	0.8
Total		120	100.0

The municipality of Addis Ababa claims that 65% of the solid waste generated is collected. Survey results indicate, however a lower coverage suggesting either over statement by the municipality or a mismatch between the perception of the municipality and households as to what constitutes acceptable service or due to the skewed service provided by the municipality. As a result of the low coverage, households disposed solid waste over a variety of sites with-in the neighborhood. These include throwing the waste in to the street, on burial sites or in riverbanks.

According to survey analysis the main environmental problem of the households is the inadequate disposal of human excreta followed by inadequate disposal of residential wastewater. The results of the analysis are listed in the following **Table 4.7**.

Environmental impacts

Human health Risks

There are some health risks associated with solid waste handling and disposal in the study area. The main problems can be classified into the following categories:

1. Presence of human fecal matter

2. The decomposition of solids into constitute chemicals which contaminates air and water systems and

3. The air pollution caused by consistently burning solid waste.

Due to the shortage of latrine services, out of the sampled households surveyed 34.2% responded their main problem is inadequate disposal of human excreta, in the study area human fecal matter is present in the solid waste management system. This presents a potential health problem to the waste management workers and to scavengers, and even community, especially children near to collection points. The usual disease pathways include placing contaminated hands in the mouth or eating food, or by directly inhaling airborne dust particles contaminated with pollutants.

Table 4.7 Main problem of the household

		Frequency	Percent	
	Inadequate disposal of human excreta	41	34.2	

	Inadequate disposal of residential wastewater	19	15.96	
Valid	Drinking water shortage	15	12.6	
	Flooding and inadequate drainage of storm water	13	10.8	
	Lack of public transport	13	10.92	
	Inadequate solid waste collection service	11	9.24	
	Presence of litter and illegal piles of solid waste	4	3.3	
	Water quality problem	2	1.7	
	Poor access for motor vehicles	1	0.8	
	Total	119	99.2	
Missing	System	1	0.8	
Total		120	100.0	

Environmental issues

The thorough observation of the study area showed that the disposal of waste in open spaces, in plantation sites, as well as in some burial places and river banks decreased the aesthetic value of those areas .The decomposition of waste into constituent chemicals is also a common source of local environmental pollution.

Plate 1. Disposal of solid waste in plantation site

Since the collected waste are disposed in the only open disposal site *'Repie'* it is wise to see the environmental issues concerned with this open disposal site. The major environmental concern is gas released by decomposing garbage. Methane is a byproduct of anaerobic respiration of bacteria, and these bacteria thrive on landfills with high amount of moisture. Methane concentration can reach up to 50 % of the composition of landfill gas at maximum anaerobic decomposition (Cointreau-Levine, 1996). In well-designed and well-sited landfills there is a potential for methane recovery; few landfills in the developing world are designed to capture and make use of methane (UNEP, 1996). Such is the case with the Addis Ababa city solid waste disposal site.

Generally, the required capital for methane recovery installation is lacking, and the low price of commercially produced gas does not make methane recovery an economically viable enterprise. Carbon dioxide is a second predominant gas emitted by landfills. Although less reactive, carbon dioxide build up in neighborhoods could be a cause of asphyxiation (Olar Zerbock, 2003).

The second problem with these gases is their contribution to the so-called green house gases (GHGs), which are blamed for global warming. Both gases are major constituents of the world's problem of GHGs; however while CO_2 is readily absorbed for use in photosynthesis; methane is less easily broken down, and is considered 20 times more potent as GHG (Johannessen, 1999).

Hoornweg, et al (1999) states that for every metric ton of unsorted municipal solid waste, 0.2 Mt are converted to land fill gases. Of these gases, CO_2 and CH_4 each comprise 0.09 Mt, since it is believed that landfill gases supply 50% of human- caused methane emission and 2.4 % of all world wide greenhouse gases (Johannessen, 1999), this is clearly an area of concern in global environmental issues.

4.2.1 Municipal service

As shown from **Table 2.3** at the municipality level there are 500 and 476 collection containers with a capacity of $8m^3$ and 1.1 m^3 respectively. If evenly distributed each sub-city will not have more than $50(8m^3)$ and $47(1.1m^3)$ communal containers. If all the population of the city uses proper collection system, a single container will be shared by up to 6070 people. This figure is three times that recommended by NUPI, i.e. one container for 2000 people.

The local and international standards set to control the location of containers shows that the distance between containers should not exceed 200m (UNHCS at NUPI, 1998). According to the Nur consult study conducted in 1982; the recommended distance between communal containers is 50 – 100 m. It seems Nur consult had accounted for the topography of the city.

Observations at the study area show that in addition to the insufficient number, the placement of the communal containers is not appropriate and also user population size is not given due consideration. In kebele 01/02 there are four $8m^3$ communal containers and all of them kept in one site situated with-in the market place. And most of the users of these communal containers are the merchants at the market place rather than households they are intended for.

37

Plate 2. Communal containers in market place at kebele 01/02

As seen from field observation and survey analysis, the long distance required to cover to dispose their waste to communal containers seems the main cause for households and some of the primary collection crews to dispose their waste in the nearby open spaces, riverbanks, burial places, plantation areas, etc.

The other kebele surveyed 03/04 have four transfer points: three of which have two communal containers each and the rest is an open pile. As compared to its adjacent kebele, the sitting of the transfer station is better but due to the undulating nature and poor settlement, the problem is still as bad as kebele 01/02.

Regarding the collection of communal containers and wastes from open piles, the Yeka sub-city has a responsibility to deliver its trucks. As seen from **Table 2.3**, the sub-city deploys eight trucks but the availability of these trucks is not always reliable due to frequent maintenance problems. Current reports revealed that only five of the trucks are functional. Survey analysis, field observation and opinion of the experts at both kebeles showed that the collection system is not functioning regularly and sometimes the garbage stays more than a week before being disposed at the final destination.

4.3 Solid waste generation rate

Globally, per capital amounts of municipal solid waste generated on a daily basis varies significantly. Economic standing is one primary determinant of how much solid waste a city produces. (World Resources institute, 1996).

In the correlation analysis performed on the data, both the sign and the relative magnitude of the correlation coefficients conformed to expectations. The household daily solid waste was positively correlated with the income (r_{xy}=0.352). This means that households with higher incomes generated larger quantities of solid waste per day.

Table 4.8 Generation rate vs. income

Income status	Generation rate (gm/cap/day)	Density (gm/cm^3)
Low income	181.13	0.328
Middle income	238.47	0.271
High income	322.36	0.211

The statistical analysis result of the 120 sampled households listed in the following **Table 4.9**.

Table 4.9 Statistical analysis results

Mean	196.84	
Std. Error of Mean	9.7904	
Median	174.00	
Mode	122.70	
Std. Deviation	105.44	

Skewness	2.962	
Std. Error of Skewness	0.225	
Kurtosis	13.782	

Since most of the previous studies made on sampling from disposal containers, it is difficult to compare with this analysis. However the study made by Yitayale (2005), which was made by sampling 197 households for 7 consecutive days, shows significant variations (100gm/cap/day). The researcher thinks that the significant variation of results occurred not only because of socio-economic variation of the studied area and study duration but more significantly due to management of data quality. With minimal project budget and with no logistic supply it would be difficult to collect and manage samples from 26 kebeles at once.

The educational level of the family head was negatively correlated with the quantity of daily household solid waste (r_{xy}=-0.31). This indicates that households whose head was highly educated generated less solid waste each day, reflecting the fact that individuals with higher level of formal education exhibit higher levels of general environmental and health awareness.

When computing the weekly and monthly generation rate per capital the results would be: 1.372 Kg/cap/week and 5.88 Kg/cap/month, respectively.

The total solid waste generated from households of the studied area can be calculated by multiplying the total population and it would be: 13,073Kg /population/day.

The weekly generation of the study area is: 91,517Kg /population/week

When taking the average density of 311 kg/m^3 the total volume of waste expected in one -week time is around 294.2 m^3. Currently the potential of the sub city to collect the wastes from the studied area is not more than 130m^3 per week taking in to account that there are 10 collection containers with the capacity of 8 m³ and from each of the five collection points a truck with a collection capacity of 10 m³ collect wastes once a week. Therefore, the collection efficiency of the sub-city around the study area is not more than 44%.

4.4 Solid waste compositions

Although countries sometimes use different categories for the physical characterization of solid waste, the categories listed in **Table 4.10** can usually be distinguished in the various waste characterization studies. Not only wealth, but also consumer patterns significantly influences waste composition (C. Zurbrugg, 2002).

According to survey analysis the following percentage composition and moisture content by weight of the different types of waste was found.

Table 4.10 Compositional analysis

Waste components	Composition by weight (%)	Moisture content (%)
Food waste	55.35	66
Ash and dirt	22.29	4.95
Yard waste	15.29	26.23
Paper	2.17	6.89
Plastic	1.60	1.62
Wood	0.92	14.71
Textile	0.83	4.95
Glass	0.67	0.059
Cardboard	0.63	7.55
Metal	0.17	1.62
Rubber	0.034	0.37

The above compositional analysis shows that the proportion of food waste takes the largest proportion and this is similar to many developing countries. The ash and dirt proportion of the domestic waste of the area is also high and this is seemingly due to the use of charcoal and wood as a major source of energy. This is directly related to the poor socio-economic condition of the households to utilize other energy sources like electricity as well as due to the poor housing conditions of the households.

Since the socioeconomic condition of the households surveyed lies more than 70.58 % in low-income range it is expected that they will have less paper and plastic wastes. The analysis also showed that the generation rate of these components increase with the income of the households. The middle-income groups generate the highest level of plastic wastes followed by the high and low-income groups. In the case of paper waste the high-income groups generate the highest.

The other waste show a lower proportion and this is likely related to the re-use behavior of the households together with selling of valuable materials. This might have resulted in decrease of waste from this category entering the waste stream.

As seen from the table the high content of food waste results in high moisture content consequently with high waste density. These physical characteristics significantly influence the feasibility of certain treatment options. Vehicles and systems operating well with low-density waste such as in industrialized countries will not be suitable or reliable under such conditions. Additionally to the extra weight, abrasiveness of the inert material such as sand and stones, and the corrosiveness caused by the high water content, causes rapid deterioration of equipments.

According to Cointreau (1982) the higher solid waste density also has many implications for the traditional method of collection and disposal. Collection and transfer trucks which are able to achieve compression rate of up to 4:1 in industrialized nations may achieve only 1.5:1 in developing countries, and land fill compression technology with average volume reduction of up to 6:1 in industrialized nations may only achieve 2:1 compaction with these increased waste densities. To this end compacter trucks therefore probably not useful in many applications. As income level of the society increases and the amount of post consumer waste such as packaging increases correspondingly, such technologies may be more appropriate.

The high moisture content and organic composition of waste may lead to problems of increased decomposition rates in the area because of the high average daily temperature. The high rainfall would only compound these problems, presenting additional challenges with proliferation of insect population and conditions conducive to propagation of diseases. To mitigate these problems, much more frequent collection is needed to remove organic waste before they are able to decompose. Although daily collection has proven unreliable or unworkable in many cities (Cointreau, 1982) perhaps a twice-weekly collection of organic material would be sufficient to reduce decomposition.

The city's solid waste management problems are different than those found in high income countries; indeed the very composition of our waste is different from that of developed nations.

Table 4.11 Waste compositions of selected cities (by percentage of weight) in industrialized, middle income, and low-income countries

Waste material	Industrialized		Middle Income		Low Income		
	Brookly n NY	London, Englan d	Medellin, Colombi a	Lagos, Nigeri a	Jakarta, Indonesi a	Karachi, Pakista n	Lucknow , India
Paper	35	37	22	14	2	<1	2
Glass	9	8	2	3	<1	<1	6
Metals	13	8	1	4	4	<1	3
Plastics	10	2	5	-	3	-	4
Leather, rubber	-	-	-	-	-	<1	-
Textiles	4	2	4	-	1	1	3
Wood, bones, straw	4	-	-	-	4	1	<1
Non-food total	74	57	34	21	15	4	18
Vegetative (putrescible)	22	28	56	60	82	56	80
Miscellaneou s inert	4	15	10	19	3	40	2

Compostable Total	26	43	66	79	85	96	82
Total	**100**	**100**	**100**	**100**	**100**	**100**	**100**

Source: Cointreau 1982

From the above table it was noted that there are several common differences in the composition of solid waste in the developing countries and some of these include:

- Waste density 2-3 times greater than industrialized nations,
- Moisture content 2-3 times greater,
- Large amount of organic waste (vegetable matter, etc.),
- Large quantities of dust, dirt

These differences from industrialized nations must be recognized both in terms of the additional problems they present as well as the potential opportunities that arise from this waste composition.

4.5 Particle size

The particle size distribution analysis shows irregularity in the particle size distribution of the solid waste. Particles with a size larger than 50mm, between 50-10mm and less than 10mm are 58.9%, 31.3%, 9.8%, respectively.

4.6 Proximate analysis result

Results of proximate analysis are given in **Table 4.12**. The results of ash content show the amount of inorganic substance that would remain after burning.

Table 4.12 Results of proximate analysis

Waste component	Moisture content (%)	VCM	Ash	Fixed carbon
Food waste	66	29.2	2.65	2.15
Yard waste	26.23	55	8.48	10.29
Cardboard	7.55	74.3	7.75	10.4
Paper	6.89	80.4	4.77	7.94

Wood	14.71	73.82	3.36	8.11
Plastic	1.62	91.4	2.26	4.72
Textile	4.95	81.5	12.67	0.88

In the following **Table 4.13** typical proximate analysis data for materials found in residential wastes is presented.

As can be seen from the table below, some of the results differ from the results of the studied area. The researcher thinks the variation of the result might have occurred due to the characteristic difference of the materials.

Table 4.13 Typical proximate analysis data for materials found in residential solid waste

Type of waste	Proximate analysis % by weight			
	Moisture	Volatile matter	Fixed carbon	Ash
Food waste	70	21.4	3.6	5.0
Paper	10.2	75.9	8.4	5.4
Plastic	0.2	95.8	2.0	2.0
Textile	10.0	66.0	17.5	6.5
Yard waste	60.0	30	9.5	0.5

Source: extracted from Techobanaglous et al., 1993

4.7 Elemental (Ultimate) analysis result

The results of the ultimate analysis are used to characterize the chemical composition of the organic matter in the solid waste. They are also used to define the proper mix of the waste materials to achieve suitable C:N ratio for biological conversion processes. Data on the ultimate analysis of the individual waste components are presented in **Table 4.14**.

Table 4.14 Elemental analysis result of waste components

Waste components	Percent by weight (dry basis)		
	% Carbon	**% Nitrogen**	**C:N**
Food waste	49.17	2.11	23.30
Yard waste	42.51	2.32	18.32
Paper	47.91	0.21	228.14
Cardboard	46.13	0.20	230.65
Wood	52.08	0.47	110.80

In the following **Table 4.15** typical data on ultimate analysis of the combustible components of residential solid waste are presented.

The value of the ultimate analysis shows that the carbon content of all the waste components have higher percentage and lower nitrogen percentage compared with the typical value of domestic solid waste.

Table 4.15 Typical data on the ultimate analysis of residential SW

Waste components	Percent by weight (dry basis)	
	% Carbon	**% Nitrogen**
Food waste	48.0	2.6
Yard waste	47.8	3.4
Paper	43.5	0.3
Cardboard	44.0	0.3
Wood	49.5	0.2

Source: extracted from Techobanaglous et al., 1993

4.8 Heating value result

The energy content of the waste components was determined using a laboratory bomb calorimeter in the Addis Ababa University Chemical Engineering laboratory.

The results are listed in the following **Table 4.16.**

Table 4.16 Energy value of the different waste components

Waste component	Energy content (Btu/lb)	
	Dry	As collected
Food waste	5821	1979.14
Cardboard	6717.62	6210.439
Paper	8108	7549.35
Plastic	15610	15357.118
Yard waste	7906	5832.25
Wood	7807	6658.59
Textile	10520	9999.26

In order to determine the total energy content, the compositional analysis result is essential. The necessary computations are presented in the following **Table 4.17**.

Table 4.17 Total energy value of the sample waste

Waste component	Composition	Energy (Btu/lb) As collected	Total energy, Btu
Food waste	55.35	1979.14	109545.34
Yard waste	15.29	5832.25	89175.1
Plastic	1.60	15357.118	24571.38
Paper	2.17	7549.35	16382.08
Textile	0.83	9999.26	8299.38
Wood	0.92	6658.59	6125.9
Cardboard	0.63	6210.439	3912.57
Total			**258011.8**

The average energy content of the domestic waste as collected per lb of waste is:

$$\text{Energy content} = \frac{258011.8 Btu}{100 lb}$$
$$= 2580.11 \frac{Btu}{lb} = 6001.27 \frac{KJ}{Kg}$$

A typical value of domestic solid waste has a total heating value of 5000 Btu/lb (Techobanaglous et al., 1993). However, the determined total heating value of the sampled domestic solid waste shows almost half the results of the typical value. The high proportion of food waste associated with its high moisture content might be the cause for its lower calorific value.

In **Table 4.18** the heating values of the different solid waste components from the Arada sub-city are presented. As can be seen, the results of each component doesn't show much difference but in calculating for the total heating value the study in Arada sub-city did not take into consideration the compositional differences of the various waste components.

Table 4.18 Calorific value of domestic solid waste from Arada sub-city

Waste component	Energy content (Btu/ lb)
Textile	8079.54
Wood	7999.25
Yard waste	7457.95
Food waste	7316.22
Paper	6913.36
Cardboard	6704.11
Total	**44450.34**

Source: extracted from Yitayal (2005)

4.9 Potential Management Options

An integrated approach to waste management will have to take into account community- and regional-specific issues and needs and formulate an integrated and appropriate set of solutions unique to each context (Senkoro 2003, Schübeler 1996, UNEP 1996, de Klundert et al 2001). Solutions, which work for some areas, will be inappropriate for others. Specific environmental conditions will dictate the appropriateness of various technologies, and the level of industrialization and technical knowledge present in the countries and cities will constrain solutions. There are different management options available, but from the environmental and socio economic point of

the study area, the following approach is emphasized

4.9.1 Waste reduction

It would seem that the easiest and most effective way to reduce the amount of waste to be disposed of would be to simply produce less in the first place. The amount of waste produced, is often a function of culture and affluence. The economic development of the country results in significant increase in MSW as production becomes cheaper. An emphasis on mass production and the development of cheap consumer goods has caused quality and longevity of goods to be sacrificed in the name of lowest market price, causing people to be more likely to simply throw away and replace items instead of repairing or maintaining them.

Since more than 70.58% of the community has less disposable income to spend on anything aside from necessities, they are also much more ingenious in reusing, preventing much of their wastes from ever entering the disposal system. This of course has a positive effect on the MSW situation. However it is important to realize that as the income of the society rises, they are likely to generate more waste per person.

At the national level, there are several methods that can be employed to reduce the production of waste. These include redesign of packaging, encouraging the use of minimal disposable material necessary to achieve the desired level of safety and convenience; increasing consumer awareness of waste reduction issues; and the promotion of producer responsibility for post-consumer waste (UNEP 1996). These goals may be achieved through a variety of measures, including legislative action and the creation of market forces and economic incentives, which would drive these reforms forward; applicability of each goal and method would depend on circumstances present in each situation. National reduction strategies such as these may not be relevant to Ethiopia at present, but as the country develops over time and the per-capita income of the population rises (and the expected increase in post-consumer waste occurs), these measures would have more effect.

4.9.2 Recycling

Waste reduction can be accomplished through the increased use of source separation and subsequent material recovery and recycling. Separating waste materials at the household level occurs to some extent on all households surveyed, and prevents the most valuable and reusable materials from being discarded. Following in-home retention of valuable material, waste-pickers currently remove most valuable materials either before garbage enters the waste stream or en route, especially in the lower and middle-income households.

In the study area there is no effort made on initiating youths for this purpose. At the municipality level there are buyers of waste materials such as papers, cardboards, metals, rubbers, plastic and the like. And these buyers should be given incentives because they are helping to divert many materials out of the waste stream. If recycling materials is an economically viable undertaking, small enterprises will continue to spring up whenever there is an opportunity. The municipality should not only recognize the trade in recyclables it should embrace it. By allowing small enterprises to address the problem, jobs will be created, and landfill space would be saved. Perhaps through micro-loans or small-scale assistance, the municipality government has to support and legitimize these enterprises.

As seen from observation and information gathered from different GO's, NGO's and individuals, there are no formally established groups on the separation and recovery of valuable materials; only the local merchants called *'kuralie'* are engaged in this job.

Information gathered from different primary collectors shows that some of them like Berhane primary solid waste collection enterprise located at the Arada sub-city, started on the recovery of paper and plastics from the waste collected but due to the uneconomic prices of the recovered materials they have stopped their initiative.

The current condition on the recovery of waste is not attractive due to poor economic return and low prices of the original materials such that engaging only on the recovery has not been seen feasible due to the labor required and low prices of the recovered materials but waste segregation and recovery at the households level should be given much more emphasis and much work has to be done on creation of public awareness.

According to compositional data analysis the recyclable materials: plastic (1.6), paper (2.17), cardboard (0.63), textile (0.83), rubber (0.034), glass (0.67), and metals (0.17): constituent 6 % of the total waste generated in the study area.

Unfortunately, not all-recyclable material can be recovered, mainly because of generator indifference to and ignorance/absence of the recycling program. Contamination of potential recyclable materials also reduces recovery potential.

4.9.3 Composting

A somewhat more low-technology approach to waste reduction is composting. The compositional analysis of the waste in the area showed that there is great potential for the reduction of a large proportion of waste through composting, due to a much higher composition of organic material, which accounts for 75.19 % of the total waste.

There are many advantages to composting; first and foremost, it reduces the amount of waste requiring ultimate disposal extending the life of the existing landfill. When done correctly, the end result becomes a useful product, capable of being used at the household or farm level to augment soil nutrient levels, increase organic matter in the soil and increase soil fertility. If the product is of high quality and the market exists, then the product can be sold.

There are three levels at which composting can be implemented; the residential level, the community level and the centralized large scale (more capital investment is required). Composting would work best if it is implemented at the household level but it is also feasible at the community level. Centralized large scale composting is not feasible in the current condition due to the overall cost, financial commitment required, as well as the effort required to maintain equipment sufficiently to keep a large-scale operation running.

The community composting seems feasible, owing to government help (providing rent free land to establish the site), extensive worker and community education, and establishing a distributor for the final product long before construction began.

The sub-city administration with the respective Kebeles has to work in the initiation of the society in the establishment of a composting scheme. The accessibility of the composting area will also determines the overall achievement of the scheme since the topographic nature of the study area is up and down the site should have good accessibility and not far enough from the residential areas. This scheme may also have to engage in the production of vegetables and fruits. This is just in case if there is limited market for the processed compost the scheme will consume the product and decreases the failure of the project.

Household composting has the greatest potential, since it requires no capital investment or any additional land, for those where small scale agriculture is found with in their yard and from the observation and survey analysis in the study area there are many households with some sort of garden within their boundaries. The final product can be used for their own household plots.
Education is the key to promoting this type of project, since many people in the study area are concerned regarding possible disease, odors, and pest problems. These issues rarely occur in a properly maintained compost pile; education regarding what waste should be added and how to properly construct a compost pile to eliminate rodents, would overcome most concerns.

Composting operation

The composting operation consists of three basic steps pre-processing of the waste, decomposition of the organic fraction of the waste, and preparation and marketing of the final compost product. Receiving, removal of recoverable materials, size reduction, and the adjustment of the waste properties (addition of water and nutrients) are the essential steps in the pre processing for composting.

To accomplish the decomposition step, several techniques are available including windrow, static pile, and in-vessel composting. These process differ primarily in the method used to aerate the organic fraction of solid waste, the biological principles remains the same, and when designed and operated properly all produce a similar quality compost in approximately the same time period (Techobanaglous et al., 1993)

Critical parameters in the control of aerobic composting processes include moisture content, C:N ratio and temperature.

Design and operational considerations

The following principal design considerations associated with the biological decomposition of prepared solid waste are discussed in the following paragraph.

Particle size

As seen from the survey analysis, most materials comprising the organic fractions of domestic solid waste were irregular in shape. The irregularity can be reduced substantially by shredding the organic material before composting. Most important of all, a reduced particle size increases the biochemical reaction rate during aerobic composting process. The most desirable particle size for composting is less than 5cm, but the larger particles can be composted (Techobanaglous et al., 1993)

Carbon to Nitrogen-Ratio (C:N)

The most critical environmental factor for composting is the carbon to nitrogen ratio (C:N ratio). The optimum range for most organic waste is from 20 to 25:1(Techobanaglous et al., 1993). Past experiences showed that the rate of decomposition declines when the C:N ratio exceeds this range on the other hand, nitrogen probably will be lost at ratios lower than 20:1(UNEP,2005).

Adding a nitrogenous waste to the compost feedstock can lower the unfavorable high C:N. If economics permits, it is also possible to lower by adding a chemical nitrogen fertilizer like urea. On the contrary, a carbonaceous waste can be used to elevate a low C:N. The carbon and nitrogen content as well as the C:N ratio of the various organic waste were listed in **Table 4.14**. And the results reveled that the total C:N ratio of the sampled solid waste is around 23:1 which is in the permissible range for composting

Moisture content

The optimum moisture content for aerobic composting is the range of 50-60 %(Techobanaglous et al., 1993). The compositional analysis showed that the organic portion of the waste that can easily be decomposable has a moisture content of around 54.89% and the waste from the study area is within the range of feasibility. However, if collection is delayed, it may lose some of its moisture. The moisture can be adjusted by blending of components or by addition of water.

Blending

Two design factors that may affect the blending of organic fraction of municipal solid wastes for composting are C:N ratio and moisture content. Since the compositional analysis of the residential solid waste shows a large proportion of food and yard waste the problem encountered with C:N ratio is minimal. Similarly, materials too wet or too dry for good composting can be blended in proper proportion to achieve optimum moisture content.

Mixing/turning

Initial mixing of organic waste is essential to increase or decrease the moisture content to an optimum level. For an organic waste having maximum moisture of 55-60% and a composting period of 15 days, the first turn has to be suggested at the third day (Techobanaglous et al., 1993).

Although the ultimate reason for the turning process is for the accomplishment of aeration, turning does simultaneously other beneficial functions. In periodically exposes all parts of the composting mass to the interior of the pile, i.e. to the zone of highly active microbial activity. It also may further reduce the particle size.

UNEP, 2005 outlined some key factors that should be kept in mind when piles are to be turned manually.

1. The height of the pile should not exceed that of the typical laborer.

2. Sufficient space must be incorporated in the design such that a new pile can be formed in the process of aeration.

3. During rebuilding of the pile, material from the outside layers of the original pile should be carefully placed in the interior of the newly formed pile. Since it is not always convenient to turn the

pile in such manner, in practice, supervisors should aim at trying to place material from the exterior of the pile in the interior of the new piles as often as possible during the course of the composting process. If this ideal situation cannot be achieved the deficiency can be compensated by increasing the frequency of turning (e.g., from two times per week to three times per week).

4. The new pile should be reconstructed such that the composting material is not compacted as to impede some air circulation.

Control of odor

The majority of the odor problems in aerobic composting processes are associated with the development of anaerobic conditions with in the compost pile (Techobanaglous et al., 1993). As seen from the compositional analysis, the organic portion of the waste consists of paper, plastic and similar materials. These materials normally cannot be decomposed in a relatively short period of time in the compost pile. Furthermore, because sufficient oxygen may not reach the center of such materials, anaerobic conditions can develop. Under anaerobic condition, organic acid will be produced, extremely odorous (Techobanaglous et al., 1993). To minimize the potential odor problem, it is important to reduce the particle size and remove plastics and other non-biodegradable materials from the organic material to be composted.

Economic Evaluation of compost

Cost assessment was done to determine the competitiveness of compost with artificial fertilizers. The most common fertilizers utilized in our country are urea and DAP. The nutrient compositions of these fertilizers are compared with that of compost. Data for comparison was adapted from Morarka (2006). Next, based on performance equivalence, the cost of compost was set to derive the economic potential of producing compost.

Cost of Chemical Fertilizer Use

chemical fertilizers are used for providing nutrients such as nitrogen and phosphorous. The cost of chemical fertilizer as a function of fertilizer efficiency/ performance is estimated as follows.

Urea

- Basis 100 kg of urea is used per hectare, containing 46% nitrogen (i.e., 46 Kg).
- Cost of urea: Birr.3.80 per kg and thus cost of nitrogen in the urea is Birr. 8.26 per kg.
- Plant uptake (use efficiency) ranges from 15-40 percent for nitrogen, but average of 20 % assumed.
- The actual utilized nitrogen is 9.2 Kg

DAP

- Basis 100 kg of DAP is used per hectare.
- The fertilizer contains 18 percent nitrogen (18 Kg) and 46 percent phosphorus (46Kg).
- Cost of DAP is Birr 3.30 kg and thus the cost of nutrients (combined for nitrogen plus phosphorus) is Birr 5.16/ kg.
- Plant uptake (use efficiency) is 15-40 percent for nitrogen and 10-25 percent for phosphorus. The average uptake efficiency for nitrogen and phosphorus are 20% and 15 %, respectively.
- 6.9 Kg of phosphorus and 3.6 Kg of nitrogen actively utilized

Cost of compost Use

In order to penetrate the market, the finished composted product should have equal performance with less or equal price including transportation cost.

According to UNEP, 2005 the average nutrients contents reported for compost are nitrogen ranges from 1 to 4 percent (average 2%) and phosphorus ranges from 1 to 4 % (average 2%) When municipal refuse is composted.

In addition compost contains all micronutrients and trace elements that would also add up to at least one percent equivalent of nutrients.

Basis 1000 kg of compost used on per hectare of land. This gives at least 20 Kg of nitrogen i.e., 40 Kg as 1000 Kg of compost and at an average plant uptake of 65% it will provide 26 Kg of nutrients.

In addition to nutrients, compost will also provide better aeration, water retention capacity and many other benefits. Some of the major advantages of compost use are the lower cost of labor (saving due to lesser weeds in the field) and saving from the cost of treatment for insects.

Compost on subsequent use has been found to provide at least 20-30 percent more nutrients. This ability can continuously reduce the quantities of compost used in the field over long durations (UNEP, 2005).

The estimates shows that 1000 kg of compost would have1.4 times fertilizer performance as that of 100 Kg urea and 1.88 times as that of DAP. Thus the cost of 1000 Kg of compost can be estimated using least factor: 1.4 * 3.8 * 100 = 532 Birr / tone, assume transportation cost of 100 Kg of packed compost is 10 Birr around Addis Ababa the price of the compost will be 432 Birr/tone.

The amount of solid waste generated estimated to be about 13,013 kg/day for the two kebeles .The compositional analysis result shows that 75.19% of the waste stream was organic origin. Thus, the total amount of solid waste available for composting would be about 9,830Kg/day.

According to integrated bio-farm enterprise summery report (May 2007) only 50% of the original waste converted in to finished compost because of loss during decomposition process, screening, evaporation and etc. Such that 4,915 Kg/day or 1794 tone of finished compost is expect for marketing per year. As shown earlier, one tone would at least be sold to 432 Birr. Hence, the annual sale will be 775,008 for the two studied kebeles.

Estimated cost of compost production

The major cost of composting is the land cost but since the scheme has tremendous contribution for municipal solid waste management through reducing the amount of solid waste going to land fill and it also creates job opportunities The land may be obtained for free or very low cost.
The other costs include labor cost, equipment cost, water bill cost and construction cost.

Table 4.19 Estimated capital costs for the construction of composting scheme

Item	Estimated cost
Equipment cost o Wheel- barrow o Spade o Three finger hoe o Shade plastic o Wood o Soil sieve wire o Packing plastic	15,000
Construction cost (3m*8m storage house) o Corrugated iron sheet o Nail o Wood o Labor cost	6400
Land preparation cost	1000
Mechanical turner*	250,000
Total	**272,400**

* Extracted from UNEP (2005)

The construction cost includes the construction of a shelter for the storage of the finished product and equipments.

Table 4.20 Estimated operational costs for a composting scheme

Item	Estimated cost (Birr/month)
Labour cost o Collection, sorting of organic waste o Turning of compost pile and o Processing of product	7,200 (600/month)
Water bill cost	200
Total	7,400
Contingency (10%)	740
Grand total	**8,140**

The overall operational costs of the scheme takes into account 12 men can perform the job. Therefore, to establish a composting scheme in the study area a total of 272,400 for capital investment cost and 8,140 Birr/month or 97,680 Birr/year running cost required. Compared to the total revenue that can be gained from the sale of the finished compost if a mechanical turner is incorporated the total net income of the scheme would be Birr 404,928 and also labor cost reduced, otherwise with a manual turning the net income would be Birr 654,928 taking into account that the entire finished product marketed.

As seen from other developing countries experience the municipality has to allocate 20-50 percent of its budget on solid waste management and out of which around 70% spend on collection of solid waste and disposal. However diverting 75% of the solid waste to produce compost would significantly reduce management cost and prolongs the lifetime of landfill.

4.9.4 Incineration

Another option for waste reduction and disposal is incineration. Incineration cannot be considered as a disposal option since following incineration there are still some quantity of ash to be disposed off, as well as the dispersal of some ash and constituent chemicals into the atmosphere. As seen from the proximate analysis if incineration applied more than 90% of the waste can be reduced in weight. This appears to be an extremely attractive option, but incineration is an inappropriate technology for the area under investigation in particular or even at the municipality level in general. Above all, the high financial start up and operational capital required to implement an incineration facility is the major barrier to the successful adoption. Large portions of the costs are the environmental hazard mitigation components, including emission 'scrubbers'. In addition, specific technical expertise and related general repair and maintenance technology are absent in an Ethiopia context. High cost and environmental problems have led to incinerators being shut down in many cities, among them Buenos Aires, Mexico City, Sao Paolo and New Delhi (UNEP 1996).

One of the important factors which make this option also inappropriate to the city of Addis Ababa is the high moisture content of the waste due to the high proportion of food related waste; it increases the cost of start up.

The negative consequences of incineration mostly revolve around airborne emissions. Certainly, incineration volatilizes many compounds potentially harmful to human health; metals (especially lead and mercury), organics (dioxins), acid gases (sulpher dioxide and hydrogen chloride), nitrogen oxides, as well as carbon monoxide and dust (UNEP, 1996)

Thus, waste reduction through incineration is not feasible from an economic, technical or environmental point of view in the area under investigation due to the existing waste characteristics and economic background.

4.9.5 Disposal

There is always waste which needs to be disposed. Planning and execution of proper solid waste disposal is crucial in the management of solid waste. As seen from observation and survey analysis more than 58% of the household received a primary collection service such that strengthening this scheme should be given proper attention. However, the observation reveals that the schemes are running with a traditional way of collecting garbage from households. The equipments they are using are incompatible with the topographic nature of the area, no proper containers for storing wastes, no segregation of wastes at the household level together with inappropriate and insufficient placement of communal containers, makes the work of waste collection schemes more difficult.

The rest of the households surveyed disposed their waste either in the communal container or simply in the near by vacant spaces. Due to the inappropriate and insufficient placement of the communal containers it was observed that the households prefer throwing their wastes in the near by vacant spaces and river banks instead of walking a long distance to dispose of their waste in the communal container.

Priority has to be given to neighborhoods with lower economic status since they can not afford the payment for primary collection schemes. Such arrangements would encourage them to dispose their waste to nearby communal container.

As part of the solid waste management it is apparent to have a transfer station and this will open up the possibility of establishing a combined transfer station/resource recovery facility. The main objective of the resource recovery portion of the facility would be to recover materials that would be used near the facility (i.e. recyclable materials and organic waste). Therefore, only those materials that have no market or use would have to be transported for disposal. The implementation of a system of this type could lead to additional cost savings since less waste would ultimately be hauled from the transfer station to the disposal site.

Ideally, a transfer station should be located such that unit cost is minimized as a function of travel time of the collection carts to the transfer station and the time required for the transfer vehicles to travel from the transfer station to the disposal site.

Optimum number and location of transfer stations can be researched or determined by different research techniques. In practice, only a limited number of sites are feasible, due to a number of factors such as access, topography, cost and environmental acceptability.

Taking into account the above listed constraints for the implementation of transfer stations, this study recommends positioning beside the placement of communal containers especially, in areas where people with lower economic conditions live. It is apparent to have 3 transfer stations at the kebele 01/02. However, at kebele 03/04 the transfer stations are sufficient taking into account the above constraints. What is needed is making the collection scheme more efficient by placing communal containers at least in areas where the primary collection schemes do not reach.

V. CONCLUSIONS AND RECOMMENDATION

5.1 CONCLUSION

During the study it was learnt that the area studied is facing environmental degradation and public health risk due to uncollected disposal of waste on the streets, burial areas and other public areas including the market places, drainage congestions by indiscriminately dumped waste and contamination of water resources near uncontrolled dumping sites.

The survey analysis shows that the placement of communal containers is not appropriate and sufficient and this discourages peoples from disposing into the container.

On the other hand, the responsible body for management of solid waste lacks the technical and financial resources to safely manage solid waste and create adequate provisions for storing the waste at the point of generation as well as efficient and sufficient collection services. Final disposal is usually a matter of transporting the collected waste to the open dumpsite and then discharging them.

However, important progress has been made in the waste management sector over the last few years. The most important improvement is the increased level of awareness among both the public and politicians.

Based on the results of the analysis, the population in the study area generates on average a total of 13,073 kg of solid waste per day; the average rate per person is 0.196 kg of solid waste per day.

The distribution of daily household solid waste by type was: food waste 55.35%, yard waste, 15.29%, ash and dirt 22.29% and other items which are recyclable amounts to 6.1%.

As expected, a number of household related factors affected the daily quantity of solid waste generated by families. These include socio-economic and education of the head (to some extent).

62

The fact that the educational level of the family head was negatively associated with the daily quantity solid waste indicates that improving the general public's awareness concerning the problem of urban solid waste should be a high priority for the responsible policy makers.

The proximate analysis revealed high moisture content of the waste and this directly related to the high consumption of fresh vegetables and fruits. As a consequence of this high moisture content and large proportion of food waste, which is around 55.35%, the calorific value of the material is reduced. The computed calorific value also revealed that it is below the typical value for domestic solid waste and this makes incineration an inappropriate management option together with its high initial investment, maintenance costs, skilled manpower requirement and the environmental burdens.

The SWM hierarchy advocated in many literatures was found inappropriate taking into account the existing characteristics of solid waste. Rather, the first priority should rest in identifying methods to divert organics from entering the MSW stream, which then requires organized collection and other forms of management. The reason is that organics are usually the largest component of MSW and the greatest reduction in waste for collection and disposal can be achieved by diverting this component of the waste stream.

The organic fraction of the waste is the important component, not only because it constitutes a sizable fraction of the solid waste stream but also because of its potentially adverse impact up on the public health and environmental quality. A major adverse impact is its attraction of rodents and vector insects for which it provides food and shelter. Impact on environmental quality takes the form of foul odors and unsightliness. Unless organic waste is appropriately managed, its adverse impact will continue until it has fully decomposed or otherwise stabilized.

To this end, this study showed that composting together with recycling can alleviate the problems of domestic solid wastes which constituent much of the proportion that degrades the environment. In addition, the environmental, economic and social assessments also revealed that these management options are feasible in the existing environmental and social conditions.

Periodic research is needed to update and modify the data and information regarding household solid waste at the kebele and city level and planning solid waste management accordingly.

5.2 RECOMMENDATIONS

The study shows that though various initiative are undertaken to alleviate the problem of solid waste, the service still falls short of the required level. The following points are outlined for the improvement of the service.

- ❖ For efficient collection of solid waste, the availability of containers is crucial as was seen from the study, where there were a total of 10 communal containers for a population of around 66703 (1:6670). Assuming that 2000 people can use one container, then a total of 34 are actually required.

- ❖ The location of the existing containers should be changed to the center of the user population. Containers should not be sited in market places, along roads or rivers. Areas for larger containers should be delineated, protected and fenced as well as the ground should being built of concrete slab.

- ❖ During the transportation of the communal containers, the containers should have cover so that falling of waste can be reduced.

- ❖ The existing trip of trucks allows the collection of only 44%. Considering the expected amount of waste generated per week from the study area is 91.51 tones.

- ❖ The sitting of the composting area should not be far from the residential houses, assessable for trucks, not have more than 5% slope to reduce the leaching of the materials and should be fenced. The composting scheme should also incorporate supporting activities for sustainability, such as the production of vegetables as well as recycling.

- ❖ Improving the existing SWM requires a coordinated and collective effort. The overall initiative by the municipality shows the political will of the government. Regular campaign, education, and training programs should be carried out at the grass root level to create public awareness.

❖ NGO's working in this area should work together with the responsible authority in the planning and execution of their programs so that the collective effort can lead to a sustainable improvement in solid waste management.

❖ Improving the operational and management capabilities of individuals and institutions involved in MSWM at the local level is extremely important. Since the study reveals that at sub-city as well as kebele level there is a lack of trained manpower in the sector. The administration simply assigned personnel who haven't qualified for other positions to fill the vacancies.

❖ One of the most pressing areas for future progress is the field of information and education. There need to be a coordinated strategy with regard to information provision and considerable work needs to be done to change people's attitudes towards waste management as a whole and increases participation in recycling and minimization schemes. Once the public is participating in recycling schemes it is important to give them feedback on what happens with the recycled waste to keep up their interest in participation.

❖ Considerable investment should be made in emerging technologies and support should be given to research and development (R&D) in the area of technology transfer and testing of solid waste management in a local context.

REFERENCES

Adams, R. C.; F. S. MacLean; J. K. Dixon; F. M. Bennett; G. I. Martin; and R. C. Lough(1951). The utilization of organic waste in N.Z.: Second interim report of the inter-departmental committee.

Addis Ababa Chamber of Commerce (2002). Addis Ababa solid waste collection, Removal and utilization, Addis Ababa.

AASBPDA. (2004). Current Status of Solid Waste Management of Addis Ababa City, Addis Ababa.

AASBPDA. (2004). Sanitation, Beautification and Park Development Sectoral Plan (2004 - 2025), Addis Ababa.

AAWSSA. (1993). Master plan study for the development of wastewater facilities for the city of Addis Ababa, Ethiopia.

Arlosoroff, Saul (1982). "WB/UNDP Integrated resource recovery project: Recycling of Waste in developing countries." In: Appropriate Waste Management for Developing Countries, edited by Kriton Curi. New York: Plenum Press, 1985

Beyene Geleta. (1985). Managing solid waste in Addis Ababa, Paper presented on the 25[th]WEDC Conference on Integrated Development for Water supply and Sanitation, Addis Ababa.

Central Statistical Agency (2006). Ethiopia Statistical Abstract 2005, Addis Ababa.

Chris Zurbrugg (2003). Solid waste management in developing countries.

Christian Zurbrugg (2002). Urban solid waste management in low-income countries of Asia: how to cope with the garbage crisis. Paper presented for scientific committee on problems of environment (scope) urban solid waste management review session, Durban, South Africa.

Cointreau, S.(1982). Environmental Management of Urban Solid Waste in Developing Countries: A Project Guide. Washington, DC: Urban Development Department, World Bank.

Fitsum Haile (2003). Existing situations of waste management in urban areas of Ethiopia, ministry of Federal Affairs: National urban planning institute, urban development policy project, Addis Ababa.

George Techobanaglous, Hilary Theisen and Samuel A. Vigil (1977). Integrated solid waste Management, Mc Grow Hill, New York.

Gerard keily (1997). Environmental Engineering. Mc Grow Hill, London.

Helge Borattebq (1997). Emerging strategies and concepts – A Framework for the Future in Sustainable Industrial Production: the Baltic university programme, uppsala university, Sweden.

Hoornweg, D.; Thomas, L. and Otten, L. (1999). Composting and Its Applicability in Developing Countries. Urban Waste Management Working Paper Series 8. Washington, DC: World Bank.

Johannessen, L.M. (1999). Observations of solid waste landfills in developing countries: Africa, Asia and Latin America. Urban and Local Government Working Paper Series No. 3, The World Bank, Washington, DC.

Jorge Jaramillo (2003), Guidelines for the Design, Construction and Operation of Manual Sanitary Landfills, Universidad de Antioquia, Colombia.

Mir Anjum Altafa and J.R Deshazo (1996). Household demand for improved solid waste management: A case study of Gujranwala, Pakistan. Elsevely science Ltd.

Murray Cullen, Lecture Notes on Waste Technology [Online 2007, June 27, Web Site at http://www.scu.edu.au/staff_pages/mcullen/lecture.html].

Olar Zerbock (2003). Urban solid waste management: waste reduction in Developing Nations. Micigan technological university.

Otoniel, B. (2001). forecasting generation of urban solid waste in developing countries, A case study in Mexico, air and waste management association.

Pavoni, J., Heer, J. and Hagerty, D. (1975). Hand Book of Solid Waste Disposal: Material and Energy Recovery. Van Nostrand Reinhold Co. New York.

Peter, S. (1996). "Urban Management And Infrastructure - Conceptual Framework for Municipal Solid Waste Management in Low-Income Countries", UNDP/UNCHS (Habitat)/World Bank/SDC Collaborative Programme on Municipal Solid Waste management in Low-Income Countries.

P.A. koushki and A.L.Al-Khaleeti (1998). An analysis of household solid waste in Kuwait: Magnitude, Type, and Forecasting Models, technical paper. Department of civil engineering, Kuwait University, Safat, Kuwait.

Regulatory and Institutional reform in the Municipal solid waste management sector (2004). Ethiopia base line Review of current waste management situation in Ethiopia, Final Report,Addis Ababa.

U.S. Environmental protection Agency of solid waste (1999). Characterization of Municipal Solid waste in the United States. Report No. EPA 530.

UNEP. (1996). International Source Book on Environmentally Sound Technologies for Municipal Solid Waste Management. UNEP Technical Publication 6, Nov. 1996.

UNEP. (2005). Solid Waste Management. UNEP technical publication volume I.

Weiss, N.A. (1988). Elementary statistics. Addison-Weslesy Publishing Company.

World Resource Institute, United Nations Environmental Programme, United Nation Development Program, The World Bank, 1996, World Resources 1996-97 – The Urban Environment, Oxford University Press, Oxford.

Yami Birke(1999). Solid waste management in Ethiopia. Paper presented on the 25th WEDC conference on Integrated Development for water supply and sanitation, Addis Ababa.

Yitayal Beyene (2005). Domestic Solid Waste Quantity And Composition Analysis in Arada Sub-City, Addis Ababa, Addis Ababa University, Addis Ababa.

Zerayakob Belete (2002). Analysis and Development of Solid Waste Management system of Addis Ababa, Addis Ababa University, Addis Ababa.

ANNEX -1 SURVEY QUESTIONNAIRE

For assessing the socio-economic situation and solid waste management behavior of the population.

Date of interview:

Name of interviewer:

Kebele:

House Number:

A. Identification:

"I would like to ask you some questions that would assist in the determination of how much solid waste is generated from each individual and its composition in order to determine how effectively we can manage our solid waste. These questions usually take about 10 minutes. We are interviewing a sample of 120 households in your neighborhood, so your input is considered very valuable to this survey. Let me first ask you a few questions to identify this house and you."

A.1   Household identification:

A.2   Name of Respondent:

A.3   Position of Respondent:

    Head of household

      Spouse of head of household

       Other   , please describe ……………………………………………………………..

A.4   How many people (children and adults) live in your household on a regular basis? …………..

B. Major Concerns:

(For this question, present the list in a different order on a random basis to each respondent)
"I would like to show you a list of possible problems that might be faced by your household:
- a) Difficult access to drinking water
- b) Poor quality of drinking water
- c) Inadequate disposal of residential wastewater
- d) Inadequate disposal of human excreta
- e) Flooding and inadequate drainage of storm water

f) Poor access for motor vehicles

g) Lack of public transport

h) Unreliable electricity supply

i) Inadequate solid waste collection service

j) Presence of litter and illegal piles of solid waste

k) Nuisance from solid waste transfer points

l) Nuisance from solid waste disposal sites

B.1 Of these possible problems, which do you consider the most serious problem for your household?

Most serious problem .... (Write letter – a to l.)

Don't know

B.2. And which do you consider the second most serious problem?

Second most serious problem ....(Write letter – a to l.)

Don't know

B.3 (If item (i) was not listed) In your opinion, how serious is the problem of solid waste collection in this area?

Very serious   a

Somewhat serious   b

Not serious   c

Don't know   d

B.4 (If item (j) was not listed) In your opinion, how serious is the problem of littering and illegal piles of solid waste in this area?

Very serious   a

Somewhat serious   b

Not serious   c

Don't know   d

B.5 (If item (k) was not listed) In your opinion, how serious is the problem of nuisance from solid waste transfer points in this area?

Very serious   a

Somewhat serious   b

Not serious   c

Don't know   d

B.6 (If item (l) was not listed) In your opinion, how serious is the problem of nuisance from solid waste disposal or dumping in this area?

Very serious   a

Somewhat serious   b

Not serious   c

Don't know   d

C. Existing Situation Regarding Solid Waste:

"I would like to ask you some questions regarding the collection or removal of solid waste from your household."

C.1 Does your household have a durable metal or plastic container for storing solid waste?

Yes, we have metal or plastic container   a

We have basket or carton container   b

No, we do not have a container   c

Don't know   d

C.2 Do you reuse the solid waste?   Yes   No

If yes, what type...

C.3 what type of solid waste is common in dry and wet season?

Dry season ....

Wet season ....

C.4 Do you compost waste? Yes   No

If yes what type of solid waste...
For what purpose...

C.5 Do you sell your waste? Yes   No

What type of waste......................................

C.6 Do you use the solid waste as a source of energy? Yes   No

What type of waste..

C.7 Does your household receive a collection service of any type?

     Yes   a (Go to Question C.8)

     No   b

     Don't know   c (Try question C.8)

C.8 How frequently is your container usually taken out to be emptied?

     Several times each day   a

     Daily   b

     Three times a week   c

     Twice a week   d

     Once a week   e

     Less frequently   f

     Don't know   g

C.9 Who usually takes the container with its waste contents out to be emptied?

     Head of household (or establishment) a

Spouse of head of household (or establishment) b

Another male adult   c (Please specify) …………………………………………..

Another female adult   d (Please specify) …………………………………………

Any male adult   e

Any female adult   f

Any child between the ages of 13 and 18   g

Any child between the ages of 6 and 12   h

Don't know   i

C.10   Where is your container taken to be emptied?

The container is placed beside the road for emptying into a collection vehicle   a

The container is emptied into a larger container in the same building   b

The container is emptied into a communal container in the neighborhood.   c

The container is emptied onto an open pile of waste in the neighborhood.   d

The container is emptied at the final disposal, and the waste stays there   e

Don't know   f

C.11   Approximately how far or how many minutes walking time one-way is it to empty your container?

………. meters one-way

………. minutes walking one-way

Don't know

C.12   If your container is emptied into a larger container in the same building or into a communal container in the neighborhood, how often is that (larger) container emptied?

Daily   a

Three times a week   b

Twice a week   c

Once a week   d

Less than once a week   e

Less than once in 2 weeks   f

Less than once in 3 weeks   g

Less than once a month   h

Don't know   i

C.13   If your container is emptied onto an open pile of waste in the neighborhood, how often is that pile removed?

Daily   a

Three times a week   b

Twice a week   c

Once a week   d

Less than once a week   e

Less than once in 2 weeks   f

Less than once in 3 weeks   g

Less than once a month   h

Don't know   i

C.14   For how many years has this type of waste collection service been provided to your household?

Less than one year   a

One to two years   b

Two to five years   c

More than five years   d

Don't know   e

C.15 Who collects the waste from the curbside, communal container, or pile?

Local government   a

Local public authority   b

Neighborhood group   c

Private company   d

Don't know   e

C.16 Has the same organization been collecting the waste for the past five years, or has there been a
change in who has been collecting your waste?

The same organization for the last five years   a

There has been a change in the last five years.   b

Don't know   c

If there has been a change, please give more details ....

....

C.17 What is your opinion of the service that you are receiving for collection of solid waste from
your household?

Very satisfied   a   Go to Question C.19

Reasonably satisfied   b   Go to Question C.19

Not satisfied at all   c   Go to Question C.18

Don't know   d

C.18 If you are not satisfied with service, would you state your <u>primary</u> reason?

The service is not reliable   a

Frequency of service – the interval between collections is too long.   b

The location of the communal container or pick-up point is unsatisfactory   c

Lack of clean appearance, odors, flies or fires at the communal container.   d

The collection workers are rude or impolite.   e

Lack of clean appearance of the neighborhood   f

Other problem   g   Please explain..

....

C.19 Do you know where the collected waste is taken for final disposal when it leaves your
neighborhood?

    Yes   a   Go to Question C.20

    Don't know   b

C.20 Are you concerned about whether the final disposal is environmentally safe and acceptable?

    Yes   a

    No   b

    Don't know   c

D Socio economic situation

No	Name	sex	Age	Marital status				children			remark
				single	married	divorced	windowed	F	M	age	

E. Education and Income

Education	Income per month(for the last 12 month)	Expenditure per month		Income status			Remark
		House rent	Other expenses	High	Medium	Low	

F. Housing characteristics

House ownership				Purpose of the house		Remark
Owned	Rent from			Living only	Living + other purpose	
	Individual	Kebele	Housing			

G. Types of energy source

Frequency of use	Types of energy source							
	Electricity	Gas	kerosene	charcoal	Firewood	Dung/Manure	Paper	plastic
Most of the time								
Some time								
Rarely								

H . Other Information

"We will soon be ending this interview. Before we do end it, I would like to ask some questions about you and your family."

F.1 What is your age?   Under 24   a ,   25 to 34   b ,   35 to 44   c ,   45 to 54   d ,   55 to 64   e ,   Over 65   f.

F.2 What is your level of education (number of years of school)? .... years

F.3 What is the level of education of the most educated member of your household?
.... years at school

F.4 How many children under 15 years of age are in your household? ....

F.5 How many people in your household contribute to the household income?

.... people

F.6 What is the occupation of the principle income earner in the household?

Self-employed as laborer a

Self-employed as trader b

Self-employed as consultant or professional c

Employee of a private company d

Employee of government (public sector) e

Retired f

Other g

Don't know h

ANNEX- 2 PLATES

Plate 1. The different color sample collection bags

Plate 2 .Preparation for sorting

Plate 3.Volume determination of the different components

Plate 4. Determination of particle size

Plate 5. While taking the data on the weight of the sample

Plate 6. The different components getting chopped

Plate 7. Equipments used for volume and particle size determination

Plate 8. Laboratory analysis

Plate 9. The crews participated in waste collection